Quantum Cosmos: The Wave Function of the Universe

1. Foundations of Quantum Cosmology...10
2. The Wheeler-DeWitt Equation...22
3. The Many-Worlds Interpretation and Cosmology...34
4. The Hartle-Hawking State...46
5. Decoherence and the Emergence of Classicality...58
6. Time and the Wave Function...71
7. Quantum Tunneling and the Birth of the Universe...84
8. Observables in a Quantum Universe...94
9. The Anthropic Principle and Wave Function Probabilities...108
10. Future Directions in Quantum Cosmology...122

Dive into the mysteries of the cosmos with *Quantum Cosmos: The Wave Function of the Universe*. This groundbreaking exploration takes you on a journey to understand the fundamental quantum framework that shapes everything we see—and much more. From the origins of quantum cosmology to the far-reaching implications of cutting-edge theories, this book unravels the secrets of the universe's wave function in an accessible yet thought-provoking way.

Begin with the *Foundations of Quantum Cosmology*, where the story of the universe is told through the lens of quantum mechanics. Learn how the *Wheeler-DeWitt Equation* redefines gravity and time, and explore the mind-bending *Many-Worlds Interpretation*, revealing a multiverse of infinite possibilities. Discover the revolutionary *Hartle-Hawking State*, proposing a universe without boundaries, and uncover how *Decoherence* bridges quantum chaos with the classical world we observe today.

What is time, really? The chapter *Time and the Wave Function* challenges everything you thought you knew. Then, delve into how *Quantum Tunneling* sparked the birth of our universe, and consider the limits of observation in a universe governed by quantum rules. The book also connects the wave function to profound questions of existence in *The Anthropic Principle* and speculates on the future of quantum cosmology with emerging theories.

Perfect for science enthusiasts and professionals alike, *Quantum Cosmos: The Wave Function of the Universe* offers a captivating exploration of the quantum foundations of reality, inviting you to see the cosmos in an entirely new light.

About Scott Perdue

Scott Perdue is a dynamic entrepreneur, author, and community leader with a life rooted in faith, family, and service. A devoted Christian, Scott has been married for over 20 years and is the proud father of four children—two girls and two boys. His passion for personal development and spiritual growth is reflected in his prolific writing career, having authored over 100 books, most of which focus on self-help and Christian themes. His books have touched the lives of countless readers seeking guidance on how to lead a fulfilling, faith-centered life.

For over 15 years, Scott has been a dedicated member of GUTS Church, a place he fondly refers to as "It Takes GUTS to Serve the Lord." His service to the church and community extends beyond attendance; he spent six years as a representative for the GUTS Food Bank, where he managed the movement of wholesale goods to help those in need. Scott also led a successful Maximized Manhood study group based on Edwin Cole's teachings, further exemplifying his commitment to fostering spiritual growth among men.

An accomplished entrepreneur, Scott has started and operated over 30 businesses, ranging from pest control to contracting. He is the founder of Universal Bug Man, a pest control service where Scott earned a reputation as a "pest control superhero." His entrepreneurial ventures include Tulsa Furniture Wholesale, Tulsa Auction Spot, and Builderhaus Unlimited, among others. Scott's business acumen extends to the health and wellness industry, where his company HCG Medical helped over 20,000 clients lose weight, generating over $6.5 million in sales in its best year.

Scott Perdue is a man of many talents, driven by his faith and dedication to serving others through his varied enterprises and writing.

Follow Scott Perdue on YouTube, Facebook & Visit UniversalWholesaleStore.com

Published Books by Scott Perdue (Buy Today on Amazon) ⤵

Christian Books by Scott Perdue:

Biblical Entrepreneur Leadership: Amplified Leverage Business Skills Book & Workbook

Biblical Men's Leadership Skills: Becoming an Amplified Christian Superstar Book & Workbook

Unleashing Biblical Manhood: Taking Ground Like a Warrior Book & Workbook

Promised Land Leadership: Leading an Army Like Joshua

Wilderness Wisdom of Moses: Timeless Life-Changing Leadership Lessons

Rules of Christianity According to Paul Book & Workbook

Provisional Miracles of Jesus: Provision through Supernatural Means Book & Workbook

Kingdom Money: Unlocking Biblical Secrets to Financial Success

The King's Highway: Lean into Jesus for Accelerated Success

Walk in the Works of the Lord: An Amplified Passion Understanding

God's River: Getting into the Kingdom Family Flow

Forgiven & Unoffendable: The Power of Walking Righteously

God is Real: Knowing the Spirit - A Journey Through Faith, Miracles, and Divine Presence

The Gift of Light: A Journey of Spiritual Growth for Life Expansion

On Fire For Jesus: Bring Plasma Energy to Your Heart Pump

Immortal DNA: Living Forever as an Eternal Spirit

Earth is God's Beach Ball: Celebrate the Legacy of Joyful Living

Faith in the Wilderness: Biblical Lessons for Strength and Spiritual Growth

Living on Purpose: A Comprehensive Guide to a Meaningful and Fulfilling Life

Praying for Others: Unlocking your God-Given Authority to Change Lives

Speaking in Tongues: Snippets of Life Improvement Code

Be Fruitful and Multiply: A Biblical Guide to Family Planning and Takes

Biblical Map of the Garden of Eden: Where does this Mysterious Garden Exist?

Methuselah: The Biblical Legacy of Noah's Grandfather

The Spirit of Jezebel in Modern Times: Acceptance vs Repentance

Love's Crossroads: The Rewards of Suffering for Love

Features of a Great Christian Camp: A Priority Spiritual Foundation

Daily Mercy: A Journey Through God's Grace Every Morning

Self Help Books by Scott Perdue:

You Are the Masterpiece: Center of the Universe Life Experience

Legacy Blueprint: How to Build a Generational Legacy

Accomplishing Greatness: 10 Legendary Skill Sets of Self-Made Millionaires

The Passive Income Playbook: 10 Game Changing Strategies to Build Wealth

Beginners Guide to Investing in the Future: Gain Wealth from Cutting Edge Sectors

Motivation for Creation: Unlocking the Spark Within

Master Productivity: Unlock your Path to Success

10 Step Productivity Plan: A Guide to Increasing Life's Results

Mindset of Productivity: A Defined Focused Journey

Mindful Love: Embracing Self Love Through Mindfulness and Compassion

Mindfulness for Personal Growth: Transform Your Life One Moment at a Time

The Ultimate Guide to Winning Friends and Influencing People: Master Communication

The Human Connection: Unlocking the Secrets to Understanding and Relating to Others

Stress Free Living: Simple Strategies for Modern Life

Mind Switch: Are you Over-Thinking Negative Thoughts?

Mastering Self-Control: Unleashing the Power of Discipline for Success in Every Aspect of Life

Rising From The Ashes: How to Rebuild When Life Falls Apart

Unlocking Secrets to Weight Loss: A Comprehensive Guide to Science, Nutrition and Wellness

Effective Diet Supplements for Weight Loss

The Body Detox Blueprint: 10 Essential Steps to Cleanse, Heal, and Revitalize Your Body

Secret 1000 Calorie Cryogenic Diet

Book Sales Formula: 10 Proven Secrets to SkyRocket Your Book Sales

Learn to Enjoy Reading: Your Ultimate Guide to Loving Books

The Ultimate Blueprint to Comedy: Your Guide to Mastering Humor and Making People Laugh

Decluttering Your Home: Take Control of Your Space, One Step at a Time

Real Estate Needs Observation: Hot to Bring Light to Entropy & Chaos

Business Books by Scott Perdue:

Legendary Business Skills: How to Think like an Entrepreneur

Seal the Deal: Mastering Sales Objections to Close Every Sale

10 Step Marketing Launch: Ultimate Guide for a Business Advertising Start Up

Email Marketing Success: 10 Ways to Master Business Email Advertising Strategy

Controlled Decent: How to Close a Business

How to Start a Business Networking Group: Learn to Organize and Motivate Business Leaders

Negotiate Like an Auctioneer: Mastering the Art of Persuasion and Control

Auction House Blueprint: How to Win Bids and Host a Successful Auction

How to Run an Antique Shop: Restoring Antique Relics to Modern Living

Secondhand Success: A Complete Guide to Running a Profitable Used Furniture Store

The Thrift Store Playbook: How to Build, Manage, and Thrive in the Resale Business

How to Start an RV Park: Your Roadmap to Success

Turn Rapids to Revenue: How to Run a Profitable River Float Business

Science Books by Scott Perdue:

Creation of Your Galactic Record: Big Bang, DNA, Creation of the Universe. Boom!

Symphony of Life: How Human DNA Plays Like Music

Quantum Cosmos: The Wave Function of the Universe

An Astronauts Heavenly Perspective: Planet, Society and Economy

Earth is the Seed of Life: A Geometric Flower of Life

Infinite Plants in Every Seed

Dodecahedron Earth: Exploring the Geometric Key to the Flower of Life

Pangaea Cracked Open: A Pre-Flood World without Oceans

Ancient Cathedral Architecture: A Language Of Semantics Lost in Time

Power Independence: DIY Guide to Building Off-Grid Energy Systems

Harvesting Heaven: The Ultimate Guide to DIY Rainwater Collection Systems

Farming Tactics for the Sahara Desert: Ultimate Gardening Guide for Arid Takes

Easy to Find Herbal Remedies

How to Build Free Energy Lighting: 10 Effective Easy to Build Free Energy Lights

Dynamic Forces: Exploring the Undeniable Power of Movement

Creative Books by Scott Perdue:

Zeppelin Airship Enterprise: The Future of Flight and Travel Reimagined

Ancient Plasma Energy Weapons Revealed: The Lost Technology of Energy Weapons

Echoes of Camelot: Unveiling the Secrets and Legends of the Knights

Secret Treasures of Rome Revealed: Explore the Ancient Architecture of Rome

The Egyptian Ankh: Secrets of Eternal Life and Ancient Wisdom

Giants, Nephilim, and the Legacy of Humanity: From Ancient Myths to Modern Mysteries

Prophecy of the Seven Suns: Exploring Parhelia in Biblical Prophecy

Epic Scavenger Hunt of Machu Picchu

Adventures of Buying an Island: Edge of Your Seat Suspense Thriller Adventure

My Neighbor is an Inventor: A Journey into Wilson's World of Innovation

Adventures of the Zoo Janitor: Growing Responsibility By Excellence

Exile's Genesis: Chronicles of the New Frontier

Relics of Oklahoma: Route 66 Treasure Hunt

The Oklahoma Waterfall Hunt

Foundations of Quantum Cosmology

The universe is a grand symphony, and at its heart lies the enigmatic wave function—a mathematical entity that encapsulates every possibility, every particle, and every force. To understand the universe's foundations, we must first confront this wave function and its profound implications. By venturing into quantum cosmology, we explore a realm where the infinitesimal scales of quantum mechanics meet the vast expanses of cosmic evolution. This fusion offers not just scientific insights but also philosophical revelations, experimental challenges, and a glimpse into the hidden patterns of reality.

The Foundations of Quantum Cosmology

Quantum cosmology seeks to describe the universe's origin and evolution through quantum principles. At its core lies the concept of the **wave function of the universe**, which defines all possible states of the cosmos. Unlike classical physics, where reality is deterministic and fixed, quantum mechanics reveals a probabilistic nature to existence. The wave function does not describe a single, definitive universe but a spectrum of possibilities—a quantum superposition of universes.

To grasp this, we must start with the question: *How did the universe begin?* Classical physics points to the Big Bang as the starting point, but this model breaks down when time and space compress into the singularity—a point of infinite density and zero volume. At such extremes, quantum mechanics takes over, and the wave function becomes the language of the cosmos. This function encodes the probabilistic potential for the universe to exist, evolve, and manifest as we observe it today.

Cosmic Waves and the Hidden Patterns of Reality

At the heart of quantum mechanics is the concept of waves—mathematical descriptions of probabilities, amplitudes, and interference. Just as an ocean's surface ripples with waves, the universe is thought to resonate with **cosmic waves**—vibrations that represent the quantum states of all particles and fields. These waves interact, amplify, and cancel out, weaving a tapestry of hidden patterns that form the structure of reality.

The relevance of these cosmic waves to quantum cosmology is profound. The wave function's oscillations encode the potential for different outcomes, such as how galaxies cluster or how fundamental constants emerge. For instance, the fluctuations in the quantum field during the universe's inflationary period—moments after the Big Bang—left imprints that we now observe as the Cosmic Microwave Background (CMB). These tiny variations in the CMB are the fingerprints of cosmic waves, revealing the seeds of structure in the universe.

In a broader sense, the hidden patterns of the universe reflect a balance between order and chaos. Quantum mechanics ensures a dynamic interplay between randomness and regularity, creating a cosmos that is predictable enough to form galaxies, stars, and life, yet uncertain enough to allow for creativity and diversity.

Cracking the Code: Making Complex Physics Accessible

Understanding the wave function and its role in the universe might seem daunting, but it can be made accessible through analogies, philosophical framing, and insights from experiments. One effective way to simplify this concept is to think of the wave function as a **cosmic code**—a set of instructions or probabilities dictating how reality unfolds.

1. Analogies and Intuition

Imagine you are standing in a vast, dark room. A single beam of light shines through a slit, creating a pattern of bright and dark bands on the wall. This is the famous double-slit experiment, where particles behave as waves, exhibiting interference patterns. Now, scale this idea to the cosmos. Instead of a single beam of light, the entire universe is a projection of quantum waves interfering with one another. These patterns, when observed on a cosmic scale, manifest as galaxies, stars, and planets.

2. Philosophical Implications

The wave function also raises profound questions about the nature of reality. Does it represent actual universes, or is it merely a mathematical tool to calculate probabilities? This debate touches on interpretations of quantum mechanics, such as the **Copenhagen interpretation** (where reality collapses upon observation) and the **Many-Worlds interpretation** (where all possibilities exist simultaneously). In the context of the universe, these interpretations force us to ask: *Is our cosmos one of many, or is it uniquely selected by an observer?*

3. Observational Insights

While the wave function itself is abstract and unobservable, its effects are imprinted in the universe's structure. The aforementioned CMB is a cosmic treasure map, revealing how quantum fluctuations shaped the early universe. Observatories like the Planck Telescope and upcoming missions continue to refine our understanding of these primordial waves, offering experimental validation for quantum cosmology.

4. Experimental Challenges

Testing quantum theories on cosmic scales remains a frontier of physics. Laboratories on Earth can probe the quantum realm, but reproducing conditions like those of the Big Bang is impossible. Instead, researchers rely on indirect evidence—gravitational waves, particle interactions, and large-scale simulations. These methods help decode the wave function's implications, bridging the gap between theory and reality.

The Relevance of Cosmic Waves

Cosmic waves are not just theoretical constructs; they are active agents shaping the universe. For instance, quantum field fluctuations during inflation stretched into macroscopic scales, forming the scaffolding for galaxy clusters. Similarly, these waves govern the behavior of fundamental particles, ensuring the stability of matter and energy.

Moreover, cosmic waves may offer clues to unifying quantum mechanics and general relativity. The wave function embodies quantum principles, while the universe's structure follows Einstein's equations of gravity. Reconciling these two frameworks—a quest often termed the search for quantum gravity—requires us to understand how cosmic waves operate across all scales.

The Hidden Patterns of the Universe

One of the most compelling aspects of quantum cosmology is its revelation of hidden patterns. These patterns are encoded in the wave function and manifest as the physical laws, constants, and structures we observe. For instance:
  • **Fine-tuning of Constants**: The precise values of constants like the speed of light or gravitational strength may emerge from quantum probabilities encoded in the wave function.
  • **Symmetries and Asymmetries**: The universe exhibits symmetry at its core, yet subtle asymmetries (like matter-

antimatter imbalance) are essential for its existence. These deviations reflect interference patterns within the cosmic wave function.

- **Fractal Structures**: On large scales, the distribution of galaxies and voids resembles fractals—self-similar patterns that repeat at different scales. This structure may arise from the interplay of cosmic waves over billions of years.

The Philosophical and Human Connection

Beyond its scientific implications, the wave function invites us to reconsider humanity's place in the cosmos. If reality is probabilistic and interconnected through quantum waves, then every particle, person, and planet is part of a grander whole. This perspective aligns with ancient philosophies that emphasize unity and interdependence, bridging the gap between science and spirituality.

Moreover, the wave function challenges the notion of certainty. In a quantum universe, the future is not fixed but shaped by probabilities. This uncertainty mirrors the human experience, where possibilities abound, and outcomes depend on choices and circumstances.

Cracking the Code: The Path Forward

To truly crack the code of the wave function, we must integrate multiple approaches. Theoretical physicists continue to refine mathematical models, while experimentalists seek new ways to observe quantum effects in cosmology. Meanwhile, philosophers and thinkers grapple with the deeper meanings of quantum reality.

Emerging technologies like quantum computing may also play a role. By simulating the universe's wave function, researchers could test hypotheses that are otherwise inaccessible. Similarly, collaborations between physics and data science could uncover patterns within massive datasets, shedding light on the universe's hidden order.

Conclusion

The wave function of the universe is not merely a scientific concept; it is a gateway to understanding existence itself. From its quantum origins to its cosmic implications, the wave function offers a unifying framework for exploring the mysteries of reality. By decoding its hidden patterns and cosmic waves, we not only gain insight into the universe but also into ourselves—a part of the grand, ever-evolving quantum symphony.

The quest to crack this code is both a scientific and philosophical journey, one that challenges us to rethink what it means to exist in a universe governed by quantum laws. And as we uncover more of the wave function's secrets, we may discover that the cosmos is not just a place we inhabit but a reflection of the infinite possibilities encoded within us.

The Wheeler-DeWitt Equation

At the heart of quantum cosmology lies a profound question: how do we reconcile the strange, probabilistic world of quantum mechanics with the deterministic, geometric framework of Einstein's general relativity? The **Wheeler-DeWitt equation**, a cornerstone of quantum gravity, offers a potential answer. It seeks to describe the **wave function of the universe**—a mathematical construct encapsulating all possible configurations of the cosmos. Cracking the code of this equation could unlock profound insights into the nature of reality, the origins of the universe, and the hidden patterns woven into the cosmic fabric.

To understand the Wheeler-DeWitt equation, we must first step back and explore the context it seeks to unify: the interplay between the quantum and the cosmic. On one side, quantum mechanics describes particles, waves, and probabilities on the smallest scales, governing the behavior of atoms and subatomic particles. On the other side, general relativity explains the structure of space, time, and gravity on the largest scales, from stars to galaxies to the universe itself. These two frameworks excel in their respective domains, but they fail to overlap harmoniously. The Wheeler-DeWitt equation represents a daring attempt to bridge this divide.

The Role of the Wheeler-DeWitt Equation

In its essence, the Wheeler-DeWitt equation is a quantum analog of the Einstein field equations in general relativity. Instead of describing a specific configuration of spacetime, it describes the **superposition of all possible configurations**. Each possible configuration corresponds to a "quantum state" of the universe, and the wave function encapsulates their probabilities.

Unlike Schrödinger's equation in conventional quantum mechanics, the Wheeler-DeWitt equation has no explicit time variable. This is because, in general relativity, time is not a fixed, universal backdrop—it is woven into the fabric of spacetime itself. When quantum mechanics is applied to the universe as a whole, the concept of time dissolves into the geometry of space. In the Wheeler-DeWitt framework, the evolution of the universe is not described as a sequence of events in time but as a spectrum of possibilities in a timeless, quantum realm.

This absence of time has profound philosophical implications. It challenges our intuitive understanding of cause and effect, raising questions about what "happens" before the Big Bang or what "drives" the universe's evolution. Instead of asking when things happen, we must ask how probabilities emerge and manifest in the universe we observe.

Cosmic Waves and the Wheeler-DeWitt Equation

The Wheeler-DeWitt equation treats the universe as a **quantum wave**—a vibration in a higher-dimensional space of possibilities. These **cosmic waves** encode the potential for different spacetime geometries and matter distributions, much like a musical chord contains a combination of frequencies. By understanding these waves, we can decode the hidden patterns that shape the universe.

For example, during the inflationary epoch—the brief, exponential expansion of the universe just after the Big Bang—quantum fluctuations in the cosmic wave function were stretched to macroscopic scales. These fluctuations became the seeds for all structures in the universe, from galaxies to stars to planets. Today, we observe their imprints in the Cosmic Microwave Background (CMB), the faint afterglow of the Big Bang. These patterns reflect the interplay of quantum mechanics and general relativity encoded in the Wheeler-DeWitt equation.

The relevance of cosmic waves goes beyond cosmological origins. They also govern the behavior of spacetime on small scales, where quantum effects become significant. For instance, near black holes or in the early universe, the Wheeler-DeWitt equation could reveal how spacetime "ripples" with quantum fluctuations, offering a glimpse into the elusive quantum nature of gravity.

The Hidden Patterns of the Universe

One of the most intriguing aspects of the Wheeler-DeWitt equation is its potential to reveal the **hidden patterns of the universe**. These patterns are not immediately visible in the macroscopic world but emerge when we examine the quantum fabric underlying spacetime.

1. Wave Function Interference

The wave function of the universe is a superposition of many possible states, and these states can interfere with one another. This interference creates intricate patterns, much like ripples on the surface of a pond. In the context of the Wheeler-DeWitt equation, these interference patterns determine the probabilities of different spacetime configurations. They also help explain why the universe appears classical and deterministic on large scales, despite its fundamentally quantum nature.

2. Fractal Geometry and Self-Similarity

The hidden patterns encoded in the wave function often exhibit self-similarity, a hallmark of fractal geometry. For instance, the distribution of galaxies and dark matter on large scales resembles a cosmic web, with filaments and voids repeating across different scales. This structure may reflect the interplay of quantum fluctuations and gravitational dynamics, as predicted by the Wheeler-DeWitt equation.

3. Symmetry Breaking

The universe is governed by symmetries, such as the uniformity of physical laws across space and time. However, these symmetries are often broken, leading to the diversity of structures we observe. The Wheeler-DeWitt equation captures these transitions, showing how small quantum fluctuations can amplify into large-scale asymmetries, such as the matter-antimatter imbalance in the early universe.

Making Complex Physics Accessible

The Wheeler-DeWitt equation is notoriously complex, involving advanced mathematics and abstract concepts. However, its essence can be conveyed through analogies and accessible explanations that highlight its relevance to both science and philosophy.

1. A Cosmic Symphony

Imagine the universe as a vast symphony, with each instrument representing a possible configuration of spacetime. The Wheeler-DeWitt equation is the sheet music, guiding how these instruments interact and harmonize. The resulting wave function is the symphony itself—a dynamic, probabilistic melody that evolves timelessly.

2. The Timeless Quantum Realm

The absence of time in the Wheeler-DeWitt equation can be compared to a frozen landscape. Each point in this landscape represents a possible universe, and the wave function describes the likelihood of finding ourselves in any particular point. This analogy helps demystify the concept of a timeless quantum realm while highlighting its philosophical depth.

3. Philosophical Reflections

The Wheeler-DeWitt equation raises profound questions about reality. Is time an illusion, emerging only in our perception? Does the universe exist in a superposition of all possibilities, with our observations "selecting" one outcome? These questions bridge the gap between science and philosophy, inviting readers to ponder the nature of existence.

Observational and Experimental Insights

While the Wheeler-DeWitt equation itself cannot be directly tested, its predictions can be compared to observations of the universe.
- **Cosmic Microwave Background**: The patterns in the CMB reflect quantum fluctuations in the early universe, offering indirect evidence for the quantum wave function described by the Wheeler-DeWitt equation.
- **Gravitational Waves**: These ripples in spacetime, detected by observatories like LIGO, may reveal quantum effects

predicted by the Wheeler-DeWitt framework, especially near black holes or during cosmic inflation.
• **Quantum Simulations**: Advances in quantum computing could enable simulations of the Wheeler-DeWitt equation, providing insights into its solutions and implications.

Cracking the Code of the Wave Function

Decoding the wave function of the universe through the Wheeler-DeWitt equation is a multi-faceted challenge, requiring collaboration between theorists, experimentalists, and philosophers. It involves:
  • Refining mathematical models to better capture the interplay of quantum mechanics and gravity.
  • Developing observational techniques to probe quantum effects on cosmological scales.
  • Engaging with philosophical questions about time, probability, and reality.

By cracking this code, we not only deepen our understanding of the universe but also uncover the profound unity underlying its quantum and cosmic realms.

Conclusion

The Wheeler-DeWitt equation is more than a mathematical curiosity—it is a window into the quantum foundation of the cosmos. By describing the wave function of the universe, it bridges the smallest scales of quantum mechanics with the largest scales of relativity, revealing the hidden patterns and cosmic waves that shape reality.

This journey is not just a scientific quest but a philosophical odyssey, challenging our understanding of time, existence, and the nature of reality. As we decode the Wheeler-DeWitt equation, we take a step closer to unraveling the mysteries of the quantum universe and our place within its timeless, infinite possibilities.

The Many-Worlds Interpretation and Cosmology

The nature of reality has always been one of humanity's most profound questions. At the quantum level, reality is far stranger than our everyday experiences suggest. The **Many-Worlds Interpretation (MWI)** of quantum mechanics provides one of the most intriguing perspectives: it posits that every quantum event spawns new, parallel realities. When applied to cosmology, the MWI suggests that the universe we observe is just one of countless branches, all of which are encoded in the **wave function of the universe**. Cracking the code of this wave function involves unraveling the mechanisms behind these parallel realities and their cosmic implications.

This idea transforms the wave function from a mathematical abstraction into a multiverse generator. By examining how the wave function relates to branching universes, we can explore deep philosophical questions, test theoretical predictions, and uncover hidden patterns that define the cosmic order.

The Many-Worlds Interpretation: A Radical Perspective

In classical physics, events unfold in a single, deterministic trajectory: if you know the current state of a system, you can predict its future state with certainty. Quantum mechanics, however, is governed by probabilities. The wave function of a system encapsulates all possible outcomes, each with a specific likelihood. Traditionally, the **Copenhagen interpretation** of quantum mechanics suggests that the act of observation "collapses" the wave function, selecting one outcome as reality.

The Many-Worlds Interpretation rejects this notion of collapse. Instead, it proposes that all outcomes encoded in the wave function are real. Every time a quantum event occurs—such as the decay of a particle or the spin of an electron being measured—the universe "branches," creating a new reality for each possible outcome. In this view, the wave function does not collapse; it evolves deterministically, encompassing an ever-growing multiverse.

The Wave Function and the Cosmos

When applied to the universe as a whole, the MWI raises profound questions:
- If the wave function of the universe encapsulates all possible states, does this mean there are countless versions of the universe branching into existence?
- What governs the probabilities of these branches, and how do they relate to the universe we observe?
- Are these parallel universes physically real, or are they just mathematical constructs?

The wave function of the universe is a solution to quantum equations that describe not just particles but the very fabric of spacetime. The **cosmic wave function** encodes all possible configurations of matter, energy, and geometry. Each branch of the multiverse represents a specific configuration, with its own physical laws, constants, and histories.

Cosmic Waves and Branching Universes

The concept of **cosmic waves** is central to understanding how the wave function leads to branching universes. These waves represent the probabilistic amplitudes of quantum states, and their interference patterns shape the evolution of the multiverse.

1. Quantum Interference and Branching

In quantum mechanics, interference occurs when waves overlap, either amplifying or canceling each other out. At the cosmic level, these interference patterns determine the likelihood of different branches emerging from the wave function.

For example, during the inflationary epoch, quantum fluctuations in the cosmic wave function gave rise to regions of spacetime with slightly different densities. These fluctuations set the stage for branching universes, each with its own unique cosmic history.

2. The Role of Observers

In the MWI, observers play a passive role, unlike in the Copenhagen interpretation. An observer does not collapse the wave function but becomes part of the branching process. For example, if an astronomer measures the spin of a distant particle, the universe splits into branches where the spin is measured as "up" and "down." The astronomer exists in both branches, each unaware of the other.

3. Cosmic Patterns and Parallel Realities

The interference patterns of the cosmic wave function reveal hidden structures within the multiverse. These patterns may explain why our universe has certain properties, such as the fine-tuning of physical constants that allow for life. Some branches may have drastically different laws of physics, while others may closely resemble our own.

The Hidden Patterns of the Universe

The Many-Worlds Interpretation offers a new lens to view the universe's hidden patterns. These patterns are not immediately visible in the macroscopic world but emerge from the quantum processes that govern the wave function.

1. Fine-Tuning and Anthropic Selection

One of the most striking features of our universe is the precise tuning of physical constants, such as the strength of gravity or the charge of the electron. If these constants were even slightly different, stars might not form, or chemistry might be impossible. The MWI suggests that this fine-tuning is a natural consequence of the multiverse: while most branches have conditions unsuitable for life, a small subset (including ours) supports complexity and observation. This is known as the **anthropic principle**.

2. Fractal Geometry of the Multiverse

The structure of the multiverse may exhibit fractal-like properties, with self-similar patterns repeating across different scales. This fractal geometry emerges from the recursive branching of universes, as encoded in the wave function. Observing fractal patterns in cosmic data, such as the distribution of galaxies, could provide indirect evidence for the multiverse.

3. Symmetry and Asymmetry

The multiverse is a realm of immense diversity, but it is governed by underlying symmetries. These symmetries dictate the probabilities of branching and the properties of resulting universes. However, slight asymmetries—such as the matter-antimatter imbalance in our universe—play a crucial role in shaping observable reality.

Making Complex Physics Accessible

The Many-Worlds Interpretation can be made accessible through analogies and clear explanations that highlight its philosophical depth and scientific relevance.

1. A Quantum Tree of Life

Imagine the universe as a tree, with the wave function as its trunk. Each quantum event causes the tree to branch, creating new realities. Just as a tree grows more complex over time, the multiverse expands with each branching event. This analogy helps illustrate the deterministic evolution of the wave function while emphasizing the interconnectedness of all branches.

2. Philosophical Implications

The MWI raises profound questions about identity, free will, and the nature of reality. If all possible versions of yourself exist in parallel universes, what does this mean for personal choice? Do you have free will, or are your actions determined by the branching of the wave function? These questions bridge the gap between science and philosophy, inviting readers to ponder their place in the multiverse.

3. Cosmic Relevance

The MWI is not just a theoretical curiosity—it has implications for understanding the origins and fate of the universe. By exploring the branching structure of the multiverse, we gain insight into phenomena such as black hole evaporation, cosmic inflation, and the nature of dark energy.

Observational and Experimental Insights

Testing the Many-Worlds Interpretation directly is challenging, as parallel universes are inherently inaccessible. However, indirect evidence and theoretical developments can shed light on its validity.

- **Cosmic Microwave Background (CMB)**: Patterns in the CMB, such as quantum fluctuations from inflation, may provide clues about the multiverse's branching structure.
- **Quantum Computing**: Advances in quantum computing could simulate the wave function of complex systems, offering insights into how branching occurs.
- **Interference Experiments**: Large-scale interference experiments, such as those involving gravitational waves, could reveal the effects of quantum branching on macroscopic scales.

Cracking the Code of the Wave Function

Decoding the wave function of the universe through the lens of the Many-Worlds Interpretation requires a multidisciplinary approach.
- **Theoretical Physics**: Refining the mathematical framework of the MWI and exploring its implications for quantum gravity.
- **Cosmology**: Analyzing observational data to identify patterns consistent with multiverse branching.
- **Philosophy**: Addressing the conceptual challenges posed by parallel realities and their implications for human experience.

As we decode the wave function, we unlock new ways of thinking about reality—revealing a universe that is not singular and static but dynamic and infinite.

Conclusion

The Many-Worlds Interpretation transforms the wave function from a mathematical tool into a map of an infinite multiverse. By embracing this perspective, we confront deep questions about existence, identity, and the nature of reality. The cosmic waves encoded in the wave function weave hidden patterns that define our universe and its place in the multiverse.

Cracking the code of the wave function is a journey into the heart of the quantum universe—a quest that challenges our understanding of reality while opening the door to endless possibilities. As we explore the Many-Worlds Interpretation, we take a step closer to understanding the grand tapestry of existence and our unique role within it.

The Hartle-Hawking State

The universe is vast, enigmatic, and deeply fascinating. One of the most intriguing questions in cosmology is: **How did the universe begin?** While classical theories point to the Big Bang as the starting point, quantum cosmology offers a deeper and more nuanced perspective. The **Hartle-Hawking state**, proposed by physicists James Hartle and Stephen Hawking, suggests that the universe did not have a definitive beginning in time. Instead, it emerged from a **quantum wave function** in a timeless, boundary-less state. This radical idea, often called the **no-boundary proposal**, challenges our understanding of origins and invites us to decode the hidden patterns embedded in the wave function of the universe.

Understanding the Hartle-Hawking state is key to cracking the code of the quantum universe. By exploring how the wave function arises in this model, we can uncover the role of **cosmic waves**, examine the hidden symmetries of the cosmos, and address profound philosophical questions about existence.

The Hartle-Hawking State and the No-Boundary Universe

In classical cosmology, the Big Bang is described as a singularity—a point of infinite density and zero volume where the laws of physics break down. This description, while compelling, leaves many questions unanswered. How could the universe emerge from such a singularity? What came before it?

The Hartle-Hawking state offers an elegant solution by eliminating the need for a singularity altogether. In this model, the universe's wave function describes a state where time and space are unified, and the boundaries of spacetime disappear. Instead of a sharp beginning, the universe arises smoothly from a quantum state—a **no-boundary condition**.

Imagine spacetime as a two-dimensional surface. In the no-boundary proposal, this surface is like a sphere with no edges or corners. There is no "before" or "outside"; the universe simply exists as a self-contained entity. The Hartle-Hawking state mathematically describes this timeless, boundary-less state using the principles of quantum mechanics.

The Wave Function of the Universe

The wave function of the universe, in the Hartle-Hawking model, encodes all possible configurations of spacetime and matter. It is a mathematical construct that describes the probability of the universe having certain properties, such as its size, shape, and energy distribution.

In this framework:

- The universe begins as a quantum fluctuation in a timeless, high-energy state.
- Cosmic waves, representing quantum probabilities, dictate how the universe evolves.
- As the universe expands and cools, these quantum fluctuations give rise to the classical structures we observe, such as galaxies and stars.

The wave function provides a bridge between the quantum and classical realms, showing how the timeless state of the no-boundary universe transitions into the observable cosmos.

Cosmic Waves in the No-Boundary Universe

Cosmic waves play a crucial role in the Hartle-Hawking model. These waves are quantum vibrations that represent probabilities within the wave function. Their interference patterns shape the evolution of the universe and reveal hidden structures in its fabric.

1. Quantum Fluctuations and Cosmic Structure

During the early moments of the universe, quantum fluctuations in the wave function created regions of slightly different densities. These fluctuations, amplified during the inflationary epoch, became the seeds for cosmic structure. Today, we see their imprints in the Cosmic Microwave Background (CMB)—the faint afterglow of the Big Bang.

2. Timeless Interference Patterns

In the no-boundary proposal, cosmic waves interfere timelessly. This means the patterns they create are not tied to a specific moment but represent the probabilities of different spacetime configurations. These interference patterns are like a fingerprint of the universe's quantum origins, revealing how the cosmos emerged from the Hartle-Hawking state.

3. The Transition to Classicality

As the universe expanded, cosmic waves transitioned from quantum states to classical structures. This process, called **decoherence**, explains how the smooth, timeless wave function gave rise to the distinct spacetime we observe.

The Hidden Patterns of the Universe

The Hartle-Hawking state reveals hidden patterns that govern the universe's structure and evolution. These patterns are encoded in the wave function and emerge through the interplay of quantum and classical processes.

1. Symmetry and Simplicity

One of the most striking features of the no-boundary universe is its symmetry. The wave function favors simple, symmetrical configurations, which explains why the universe appears isotropic and homogeneous on large scales. This simplicity is not random but arises from the mathematical properties of the Hartle-Hawking state.

2. Fractals and Self-Similarity

Despite its overall symmetry, the universe exhibits fractal-like patterns on smaller scales. The distribution of galaxies and dark matter forms a cosmic web, with self-similar structures repeating across different scales. These patterns reflect the influence of cosmic waves and quantum fluctuations encoded in the wave function.

3. Probabilities and Fine-Tuning

The wave function assigns probabilities to different configurations of the universe. Interestingly, the no-boundary proposal predicts a higher likelihood for universes that support complexity, such as the formation of galaxies and life. This fine-tuning is not the result of design but emerges naturally from the Hartle-Hawking state.

Making Complex Physics Accessible

The Hartle-Hawking state and the no-boundary universe can seem abstract and counterintuitive. However, these ideas can be made accessible through analogies and clear explanations.

1. A Timeless Ocean

Imagine the universe as a wave on a timeless ocean. The wave has no starting point—it simply arises naturally from the ocean's quantum vibrations. This analogy helps convey the idea of a no-boundary universe, where time and space emerge from a timeless state.

2. A Cosmic Canvas

The wave function of the universe can be compared to a cosmic canvas. Each brushstroke represents a possible configuration of spacetime, and the Hartle-Hawking state defines the rules that govern the painting. The final image—the observable universe—is shaped by the interference of these brushstrokes.

3. Philosophical Reflections

The Hartle-Hawking state raises profound philosophical questions. If time emerges from the wave function, what does this mean for our understanding of causality? Is the universe self-contained, with no need for external causes? These questions invite readers to reflect on the nature of existence.

Observational and Experimental Insights

While the Hartle-Hawking state is a theoretical model, its predictions can be tested indirectly through observations of the universe.

- **Cosmic Microwave Background (CMB)**: The patterns in the CMB provide a snapshot of the universe's early quantum state. Comparing these patterns with predictions from the no-boundary proposal can validate the Hartle-Hawking state.
- **Inflationary Cosmology**: The model predicts specific features in the inflationary epoch, such as the distribution of quantum fluctuations. Observing these features can provide further evidence.
- **Quantum Simulations**: Advances in quantum computing may allow us to simulate the wave function of the no-boundary universe, offering new insights into its properties.

Cracking the Code of the Wave Function

Decoding the wave function of the Hartle-Hawking state is a multidisciplinary challenge. It involves:

- **Mathematical Innovation**: Developing precise solutions to the equations governing the wave function.
- **Observational Advances**: Analyzing cosmic data to identify patterns predicted by the no-boundary model.
- **Philosophical Inquiry**: Exploring the implications of a timeless, self-contained universe.

By cracking this code, we gain a deeper understanding of the quantum origins of the universe and the hidden symmetries that shape reality.

Conclusion

The Hartle-Hawking state offers a revolutionary perspective on the universe's origins. By proposing a no-boundary condition, it eliminates the need for a singular beginning and provides a quantum foundation for spacetime. The wave function of the universe, shaped by cosmic waves, encodes the probabilities of all possible configurations, revealing hidden patterns and symmetries in the cosmic fabric.

Cracking the code of the wave function is not just a scientific endeavor—it is a journey into the heart of existence itself. As we explore the Hartle-Hawking state, we uncover the timeless principles that govern the universe and our place within its infinite possibilities.

Decoherence and the Emergence of Classicality

The mysteries of quantum mechanics and the structure of the universe are deeply interconnected. At the heart of this connection lies the phenomenon of **decoherence**, which explains how the universe transitions from its quantum origins to the observable classical reality we experience. This topic offers profound insights into how quantum states, characterized by superposition and uncertainty, evolve into the definite, deterministic world we inhabit. By exploring decoherence and its role in the emergence of classicality, we unlock the hidden patterns of the universe and gain a deeper understanding of its quantum foundation.

Decoherence is more than a technical concept in physics—it is a bridge between two seemingly distinct realities: the microscopic quantum world, where probabilities and wave functions reign, and the macroscopic classical world, where everyday objects follow clear trajectories and outcomes. Understanding this transition is key to "cracking the code" of the universe's wave function. **Cosmic waves**, or quantum fluctuations on a universal scale, are central to this transition, shaping the structure and behavior of the universe from its earliest moments.

The Quantum Origins of Reality

At its core, the universe is quantum mechanical. The **wave function of the universe** describes the probabilities of all possible states of matter, energy, and spacetime. This mathematical object is a central feature of quantum cosmology, offering a complete description of reality at the most fundamental level.

Quantum mechanics is characterized by phenomena like **superposition**, where particles exist in multiple states simultaneously, and **entanglement**, where distant particles exhibit correlations that defy classical explanations. These features dominate the early universe, when quantum fluctuations shaped its structure and evolution.

However, as the universe expanded and cooled, a profound transformation occurred: the quantum behaviors that defined its infancy gave way to classical structures like galaxies, stars, and planets. This transition, driven by decoherence, is what allows us to perceive the universe as stable and predictable.

What Is Decoherence?

Decoherence is the process by which a quantum system loses its coherence—its ability to maintain superposition states—due to interactions with its environment. In simpler terms, it is the mechanism that "collapses" the wave function, forcing the system to adopt a single, definite state.

For example, imagine a quantum particle that exists in a superposition of two locations. When this particle interacts with its environment, such as a field of photons or other particles, the superposition becomes entangled with the environment. The result is a loss of coherence: the particle no longer behaves quantum mechanically but instead appears in a single, classical location.

Decoherence explains why we do not observe quantum phenomena like superposition or entanglement in our macroscopic world. It is not that these phenomena cease to exist but that their effects are hidden by interactions with the environment, which "washes out" quantum behaviors.

Cosmic Waves and the Role of Decoherence

Cosmic waves, or quantum fluctuations, play a critical role in the early universe.

These waves represent variations in energy density and are a direct consequence of the universe's quantum nature. During the period of cosmic inflation, these fluctuations were stretched across vast scales, seeding the structure of the universe.

1. The Quantum to Classical Transition

Decoherence occurs as cosmic waves interact with the environment of the expanding universe. In the early universe, these interactions were limited due to its high energy density and rapid expansion. However, as the universe cooled and particles began to form, decoherence took hold.

Cosmic waves, initially described by quantum probabilities, transitioned into classical structures like density variations. These variations, in turn, became the seeds for galaxies, stars, and other cosmic phenomena. Without decoherence, the universe would remain in a quantum state, with no clear distinctions between structures.

2. Observable Evidence in the Cosmic Microwave Background

The Cosmic Microwave Background (CMB), the afterglow of the Big Bang, provides direct evidence of decoherence in action. The patterns observed in the CMB reflect the quantum fluctuations of the early universe, "frozen" into classical structures by decoherence. These patterns are a cosmic fingerprint of the wave function, revealing how quantum probabilities shaped the large-scale structure of the universe.

The Hidden Patterns of the Universe

Decoherence not only explains the transition from quantum to classical but also reveals hidden patterns that underlie the universe's structure. These patterns are encoded in the wave function and become observable through the effects of cosmic waves and decoherence.

1. Fractal Structures in Cosmic Distribution

One of the most striking patterns is the fractal-like distribution of matter in the universe. Galaxies and clusters form a cosmic web, with self-similar structures repeating at different scales. These patterns originate from quantum fluctuations, amplified by inflation and preserved by decoherence.

2. Symmetry Breaking and Structure Formation

The early universe was highly symmetrical, but small quantum fluctuations broke this symmetry, leading to the formation of distinct structures. Decoherence played a crucial role in preserving these fluctuations as classical features, ensuring that the universe developed complex, organized systems.

3. Entanglement and Correlation

Even as decoherence hides quantum phenomena, it leaves behind subtle correlations, or **entanglement**, between different parts of the universe. These correlations are reflected in the alignment of cosmic structures and the isotropy of the CMB, hinting at the underlying quantum nature of the universe.

Philosophical Implications of Decoherence

Decoherence raises profound philosophical questions about the nature of reality. If the wave function describes all possible states, what determines the specific reality we experience?

1. The Role of the Observer

In traditional interpretations of quantum mechanics, measurement collapses the wave function. However, in the context of the universe, there is no external observer. Decoherence provides an alternative explanation, showing that interactions with the environment effectively "measure" the quantum system, creating classical outcomes.

2. The Many-Worlds Interpretation

Decoherence is closely tied to the Many-Worlds Interpretation of quantum mechanics. In this view, every possible outcome of a quantum system exists in a separate branch of reality. Decoherence separates these branches, ensuring they do not interfere with each other. This interpretation suggests that the universe is a multiverse, with countless parallel realities.

3. Time and Causality

Decoherence challenges our understanding of time and causality. In the quantum realm, time is not a fundamental property but emerges from interactions and decoherence. This perspective aligns with theories in quantum gravity, which suggest that time arises as an emergent phenomenon.

Observational and Experimental Insights

Decoherence is not just a theoretical concept—it has been observed and tested in laboratory experiments and cosmological data.

1. Quantum Experiments

In controlled environments, scientists have demonstrated decoherence by isolating quantum systems and observing their interactions with the environment. For example, experiments with particles in superposition have shown how even small interactions cause the wave function to decohere, leading to classical behavior.

2. Cosmic Observations

The patterns in the CMB and the large-scale structure of the universe provide evidence of decoherence at a cosmic scale. These observations match predictions from quantum cosmology, confirming the role of decoherence in shaping the universe.

3. Future Technologies

Advances in quantum computing and simulation may allow us to model decoherence on a universal scale, offering new insights into the wave function and its role in the cosmos.

Cracking the Code of the Wave Function

Decoherence is a key to cracking the code of the wave function, unlocking the hidden patterns and principles that govern the universe. By studying decoherence, we bridge the gap between quantum mechanics and classical physics, uncovering the mechanisms that shape reality.

Understanding decoherence requires an interdisciplinary approach, combining physics, mathematics, and philosophy. It also invites us to reconsider fundamental questions about existence, causality, and the nature of time.

Conclusion

The transition from quantum states to classical reality is one of the most profound mysteries of the universe. Decoherence offers a compelling explanation, showing how cosmic waves and quantum fluctuations evolve into the observable structures we see today. This process reveals hidden patterns in the universe, from the fractal distribution of galaxies to the isotropy of the CMB, and raises philosophical questions about reality itself.

By exploring decoherence, we gain not only a deeper understanding of the universe's origins but also a glimpse into the quantum principles that continue to shape its evolution. Cracking the code of the wave function is not just a scientific challenge—it is a journey into the heart of existence, where quantum probabilities become the fabric of reality.

Time and the Wave Function

The concept of time has long been one of the most perplexing elements of physics and philosophy. Our intuitive sense of time as a flowing river, moving from the past to the future, shapes how we experience reality. Yet, in the realm of quantum cosmology, time itself vanishes as a fundamental entity. This paradox lies at the heart of modern efforts to understand the **wave function of the universe**, a mathematical framework that describes all possible states of existence.

In this exploration of "Time and the Wave Function," we dive into a world where time is not a given, but an emergent property of deeper, timeless laws. The absence of time in fundamental quantum cosmology and its reemergence in classical physics forces us to rethink the nature of reality itself. The interplay of **cosmic waves** and hidden patterns further deepens this understanding, as they act as the scaffolding upon which time and classicality arise. By "cracking the code" of the wave function, we uncover how the universe evolves and why time, as we know it, is both an illusion and a necessity.

The Timeless Foundations of Quantum Cosmology

In quantum mechanics, the **wave function** encapsulates the probabilities of all possible states of a system. When applied to the universe as a whole, the wave function takes on a more profound meaning. Known as the **universal wave function**, it is the ultimate mathematical description of existence, uniting all possibilities into a single, timeless entity.

In classical physics, time serves as a background parameter, ticking away independently of the events it measures. But in quantum cosmology, governed by the Wheeler-DeWitt equation, time disappears as an independent variable. The

Wheeler-DeWitt equation suggests that the universe's wave function does not evolve in time but is instead static, describing all potential states simultaneously.

This absence of time poses a philosophical and scientific puzzle: How can a timeless wave function give rise to the dynamic universe we observe, where cause and effect play out across a temporal dimension?

The Emergence of Time

The resolution to this paradox lies in the concept of **emergence**. While time does not exist as a fundamental property, it arises as an approximate concept when certain conditions are met. These conditions include:

1. **Quantum Decoherence:** In the early universe, quantum systems interacted with their environments, causing decoherence. This process "froze" quantum possibilities into definite,

classical outcomes, allowing the notion of time to emerge as a way to describe the sequence of these outcomes.

2. **Relational Time:** Instead of a universal clock, time can be understood relationally—an emergent property arising from changes in the configuration of matter and energy. For example, the ticking of a clock is meaningful only when compared to another process, such as the motion of planets or the oscillation of atoms.

3. **Thermodynamic Time:** The second law of thermodynamics, which describes the increase of entropy, provides a natural directionality to time. While the wave function of the universe is timeless, the growth of entropy gives rise to the arrow of time we observe in everyday life.

4. **Cosmic Waves and Time's Birth:** Quantum fluctuations in the early universe, or **cosmic waves**, created variations in energy density that acted as seeds for structure formation. These waves, when amplified during inflation,

provided a framework for the emergence of classical spacetime. The oscillations and patterns in these waves are deeply linked to the origins of time itself.

Cosmic Waves and Time's Hidden Patterns

Cosmic waves are quantum fluctuations that existed in the primordial universe. Despite their quantum origins, these fluctuations played a crucial role in shaping the universe's large-scale structure, from galaxies to cosmic voids.

1. Timeless Fluctuations

Initially, these waves were timeless. They represented quantum possibilities existing without any temporal order. As the universe expanded and cooled, interactions between particles introduced decoherence, effectively embedding these fluctuations into spacetime.

2. The Cosmic Microwave Background (CMB)

The CMB provides a snapshot of the universe when it was just 380,000 years old. The temperature fluctuations observed in the CMB map are remnants of cosmic waves. These fluctuations encode information about the universe's earliest moments, revealing hidden patterns that hint at the timeless quantum realm from which they arose.

3. Fractal Patterns and Self-Similarity

The distribution of matter in the universe exhibits fractal-like patterns—self-similar structures repeating at different scales. These patterns reflect the influence of cosmic waves and the timeless principles encoded in the wave function. Time emerges as we interpret the unfolding of these patterns on cosmological scales.

Philosophical Insights into Time

The absence of time in quantum cosmology challenges our most fundamental assumptions about reality. Time, as we perceive it, is deeply tied to human consciousness and experience. But if time is not fundamental, what does this say about causality, free will, and the nature of existence?

1. Time as an Illusion

Philosophers like Julian Barbour argue that time is an illusion—a construct of human perception. According to this view, the universe is a collection of "nows," or configurations of matter and energy. The flow of time is a mental construct, created by the way we perceive change.

2. Timeless Causality

Even without time, causality remains meaningful in quantum cosmology. Events can be understood as relationships between configurations, rather than sequential occurrences. This relational view of causality aligns with quantum mechanics, where entangled particles exhibit correlations independent of temporal order.

3. The Multiverse and Timelessness

In some interpretations of quantum mechanics, such as the Many-Worlds Interpretation, the wave function contains all possible realities. Each reality exists simultaneously, without a global timeline. This multiverse perspective further reinforces the idea that time is not fundamental but emergent.

Observational and Experimental Insights

While the idea of timelessness may seem abstract, it has tangible implications for cosmology and physics. Observations and experiments continue to shed light on the relationship between time, the wave function, and cosmic waves.

1. Quantum Experiments

Laboratory experiments have demonstrated phenomena like quantum entanglement, where correlations exist between particles regardless of temporal separation. These results challenge classical notions of time and causality, suggesting a deeper, timeless framework.

2. Cosmic Observations

The study of the CMB and large-scale structure provides a window into the early universe, where time as we know it had not yet emerged. Patterns in the CMB offer clues about the quantum origins of spacetime and the processes that led to the emergence of classical time.

3. Advances in Quantum Gravity

Theories of quantum gravity, such as string theory and loop quantum gravity, aim to reconcile general relativity with quantum mechanics. These frameworks often suggest that time is not a fundamental property but arises from deeper, timeless principles. Experimental progress in this field may one day confirm these ideas.

Cracking the Code of the Wave Function

To "crack the code" of the wave function is to understand the timeless principles that govern the universe. The wave function encodes all possibilities, but its interpretation requires us to move beyond classical intuitions about time and causality.

1. Mathematical Beauty

The wave function is a mathematical object of immense beauty, uniting probabilities, symmetries, and patterns into a single framework. Its timeless nature reflects the elegance of quantum mechanics, where simplicity gives rise to complexity.

2. Emergent Complexity

The emergence of time from a timeless wave function demonstrates the universe's capacity for self-organization. From quantum fluctuations to cosmic waves to classical spacetime, the universe evolves in ways that reveal hidden patterns and principles.

3. Interdisciplinary Insights

Understanding the wave function and the emergence of time requires insights from physics, philosophy, and mathematics. It also invites us to consider the role of consciousness and perception in shaping our understanding of reality.

Conclusion

The question of time's nature lies at the heart of quantum cosmology. By examining the wave function of the universe, we uncover a reality where time is not fundamental but emergent, arising from the interplay of quantum laws and cosmic waves. This realization forces us to rethink our most basic assumptions about causality, change, and existence.

Cosmic waves, as remnants of the universe's quantum beginnings, provide a bridge between timeless quantum mechanics and the dynamic world of classical physics. They encode the hidden patterns that shape the universe, offering a glimpse into the timeless principles that govern reality.

Cracking the code of the wave function is not just a scientific challenge but a journey into the deepest mysteries of existence. It is a quest to understand how the timeless universe gives rise to the temporal world we experience—a journey that reveals not only the mechanics of the cosmos but the profound beauty of its underlying order.

Quantum Tunneling and the Birth of the Universe

The story of the universe begins in a realm so extreme and mysterious that classical physics cannot describe it. At the dawn of time, quantum mechanics reigned supreme, governing the infinitesimal scales of energy and space that would give rise to everything we observe today. Among its most profound contributions is the phenomenon of *quantum tunneling*, a process that defies classical logic yet provides a plausible mechanism for the birth of the universe.

To understand how quantum tunneling might have initiated the Big Bang, we must delve into the wave function, a mathematical construct that encapsulates the probabilities of a particle's position, momentum, and other properties. It is the key to interpreting quantum behavior, and in the context of cosmology, it reveals a landscape of extraordinary possibilities.

The Role of Quantum Tunneling in the Big Bang

Quantum tunneling is a process by which particles pass through energy barriers they would otherwise be unable to surmount. In classical terms, this is impossible: an object lacking sufficient energy to overcome a barrier would remain trapped. But in the quantum world, the wave function permits probabilities that transcend such constraints. A particle, like an electron, can "borrow" energy temporarily (thanks to the Heisenberg uncertainty principle) to appear on the other side of a barrier.

Applied to cosmology, quantum tunneling offers a tantalizing explanation for the emergence of the universe from a state of "nothingness." In this context, "nothingness" refers not to a void but to a quantum vacuum—a state in which energy fluctuates unpredictably. Physicists speculate that the early universe existed as a metastable energy state, akin to a particle trapped within a potential well. For the universe to transition into a higher-energy state (the Big Bang), it needed a quantum "nudge"—a tunneling event.

From the Quantum Vacuum to the Big Bang

The quantum vacuum is not empty. It teems with ephemeral particles and waves, constantly fluctuating in and out of existence. This restless energy creates a fertile environment for quantum tunneling. Imagine a potential energy landscape where the early universe rests at a local minimum—a stable yet impermanent state. Through quantum tunneling, the universe transitions to a state of lower energy, releasing a colossal amount of energy in the process. This is the "bang" of the Big Bang.

The theoretical framework supporting this idea draws on the inflationary model of cosmology. According to this model, an incredibly brief yet exponential expansion of spacetime occurred immediately after the Big Bang. Quantum tunneling provides a mechanism for triggering this inflation, allowing the universe to escape its initial constraints and expand into the vast cosmos we observe today.

Cosmic Waves: The Fingerprints of Creation

The implications of quantum tunneling extend far beyond the Big Bang itself. One of the most significant predictions of inflationary cosmology is the existence of *cosmic waves*—ripples in spacetime that originated from quantum fluctuations during the universe's birth. These waves, known as primordial gravitational waves, stretch and compress spacetime, leaving an indelible imprint on the cosmic microwave background (CMB), the afterglow of the Big Bang.

By studying the CMB, scientists have uncovered subtle patterns that reflect the quantum nature of the universe's earliest moments. These patterns, such as temperature variations and polarization, offer clues about the conditions that prevailed during inflation. Cosmic waves serve as a bridge between quantum mechanics and cosmology, demonstrating how minute quantum fluctuations can shape the large-scale structure of the universe.

The Hidden Patterns of the Universe

The wave function doesn't just describe individual particles; it encapsulates the entire quantum field, a tapestry of probabilities that spans the universe. Hidden within this tapestry are patterns that reflect the interplay of quantum mechanics and cosmic evolution. These patterns manifest in phenomena like cosmic waves, the distribution of galaxies, and even the underlying structure of spacetime itself.

At the heart of these patterns lies a profound philosophical insight: the universe is not a static, deterministic entity. Instead, it is a dynamic system governed by probabilities and emergent behaviors. This perspective challenges classical notions of causality and invites us to rethink the nature of reality. The universe, in a sense, "chose" its current state from an ensemble of possibilities encoded in the wave function.

Making Complex Physics Accessible

To make the intricacies of quantum tunneling and cosmic evolution accessible, it helps to draw parallels between familiar concepts and abstract theories. For example:
- **Quantum Tunneling as a Shortcut**: Compare the tunneling process to taking a shortcut through a mountain instead of climbing over it. Although the mountain represents an energy barrier, quantum

mechanics allows particles to "tunnel" through, bypassing the climb.
• **Cosmic Waves as Musical Notes**: Think of cosmic waves as the harmonics of a grand cosmic symphony, where each note corresponds to a quantum fluctuation. These harmonics shaped the early universe and continue to resonate in the patterns we observe in the CMB.

By using metaphors and visual analogies, we can demystify the wave function and its implications, making them tangible for a broader audience.

Philosophical, Observational, and Experimental Insights

Quantum mechanics not only reshapes our understanding of the physical universe but also raises profound philosophical questions. For instance:
• **Why This Universe?**: If the wave function encodes multiple possibilities, why did the universe evolve as it did?

Does this imply the existence of a multiverse, where other possibilities play out?

- **The Nature of Time**: Quantum tunneling challenges our intuitive understanding of time, suggesting that events like the Big Bang are not bound by classical temporal constraints. Could time itself emerge from quantum processes?

Observational evidence continues to refine our understanding. The detection of gravitational waves by LIGO and Virgo has validated key predictions of general relativity, paving the way for future experiments to search for primordial gravitational waves. Meanwhile, advancements in particle physics, such as experiments at the Large Hadron Collider, probe the quantum fields that underpin the universe's fabric.

Relevance and Intrigue

Quantum tunneling's role in the Big Bang is more than an abstract theory; it is a window into the deepest mysteries of existence. It connects the smallest scales of quantum mechanics to the largest scales of cosmology, revealing a universe that is profoundly interconnected.

Cosmic waves, as relics of this quantum genesis, remind us that the universe's origins are written in patterns that we can decipher. By cracking the code of the wave function, we not only uncover the story of the cosmos but also gain insight into our place within it.

Ultimately, the study of quantum mechanics and cosmology is a journey of discovery—a quest to understand the hidden patterns that govern everything from the birth of the universe to the intricate structure of reality. Through this lens, the universe becomes not merely a collection of stars and galaxies but a living expression of quantum creativity, a symphony of probabilities resonating across the vastness of space and time.

Observables in a Quantum Universe

At the heart of quantum mechanics lies a fundamental paradox: while the universe is governed by the abstract and probabilistic wave function, our observations are limited to concrete, measurable quantities. In a quantum universe, the wave function encapsulates all possible realities, yet the act of measurement collapses this vast potentiality into a single outcome. How do we navigate this tension? What can we measure in a universe governed by a universal wave function? And how do these measurements connect to the grand cosmic scale, where quantum mechanics and the universe's large-scale structure intertwine?

To answer these questions, we must delve into the nature of observables—the measurable quantities that allow us to extract meaning from the wave function. Along the way, we'll uncover the hidden patterns of the cosmos, explore the role of cosmic waves in linking quantum mechanics to cosmology, and reflect on the philosophical implications of a reality shaped by measurement.

Observables: The Key to Quantum Measurement

In classical physics, measurement is straightforward: properties like position, velocity, or energy are well-defined and can be determined precisely. In quantum mechanics, however, the situation is more complex. Observables in the quantum world are represented by mathematical operators that act on the wave function. These operators yield eigenvalues—specific measurable quantities—when the system is in an eigenstate. Crucially, not all observables can be measured simultaneously with precision, a limitation known as the Heisenberg uncertainty principle.

Some of the most significant quantum observables include:

1. **Position and Momentum**: These observables are complementary, meaning that increasing precision in one leads to greater uncertainty in the other. This duality reflects the wave-particle nature of quantum entities.
2. **Energy**: Energy levels in quantum systems are quantized, with discrete values defined by the system's wave function. This quantization underpins phenomena like atomic spectra.
3. **Spin**: A purely quantum property with no classical analogue, spin is a form of intrinsic angular momentum. It's a key observable in understanding particles and their interactions.
4. **Wave Function Collapse**: Perhaps the most elusive observable is the collapse of the wave function itself—a process by which a quantum system transitions from a superposition of states to a single observed outcome.

The Universal Wave Function

The idea of a universal wave function, proposed by Hugh Everett in his many-worlds interpretation of quantum mechanics, takes these concepts to a cosmic level. According to this framework, the wave function doesn't just describe isolated systems—it encapsulates the entire universe, encompassing all possible configurations of matter, energy, and spacetime. Observables in this context are the tools that allow us to extract localized, meaningful measurements from this all-encompassing mathematical structure.

In a universe governed by a universal wave function, every measurement is a process of selection, where the act of observation identifies one reality from the vast space of possibilities. This idea raises profound questions: Are the outcomes we observe truly random, or are they determined by deeper patterns within the wave function? How do observables in our local environment connect to the universal wave function's global structure?

The Role of Observables in Cracking the Wave Function Code

Understanding observables is essential to decoding the wave function. Observables are not merely passive measurements; they shape how the universe's underlying quantum reality manifests. For instance:

1. **Quantum Superposition**: Observables determine how superposed states—where a system exists in multiple potential configurations simultaneously—are resolved into a single outcome. The famous double-slit experiment illustrates this, where the act of measurement collapses a particle's wave function, revealing either wave-like or particle-like behavior.
2. **Entanglement**: When two particles become entangled, measuring an observable on one particle instantaneously affects the other, regardless of distance. This phenomenon underscores the interconnectedness of

observables across the universe and hints at the non-locality of the wave function.

3. **Cosmic Observables**: On a larger scale, observables such as the cosmic microwave background (CMB) radiation and gravitational waves reveal how quantum fluctuations in the early universe have shaped its large-scale structure.

Cosmic Waves: Linking Quantum Mechanics to the Universe

Cosmic waves, such as the fluctuations imprinted in the CMB and the ripples in spacetime known as gravitational waves, are direct manifestations of quantum mechanics at the cosmic scale. These phenomena provide crucial observables for studying the wave function's influence on the universe.

1. **Cosmic Microwave Background (CMB)**: The CMB is the afterglow of the Big Bang, a snapshot of the universe when it was just 380,000 years old.

Embedded within the CMB are tiny fluctuations in temperature and density, which originated as quantum fluctuations during the inflationary epoch. Measuring these fluctuations allows scientists to probe the quantum properties of the early universe and the wave function that governed its evolution.

2. **Gravitational Waves**: Predicted by Einstein's general relativity and rooted in quantum mechanics, gravitational waves are ripples in the fabric of spacetime caused by massive accelerating objects. Observatories like LIGO and Virgo detect these waves, which carry information about their cosmic sources, from merging black holes to the quantum fluctuations during inflation.

Cosmic waves serve as observables that connect quantum mechanics to cosmology, providing a bridge between the wave function's abstract probabilities and the tangible structure of the universe.

The Hidden Patterns of the Universe

One of the most profound insights of quantum mechanics is that the universe is not chaotic but deeply patterned, with structure emerging from the probabilistic nature of the wave function. These hidden patterns can be seen at multiple scales:

1. **Microscopic Patterns**: At the quantum level, the wave function governs the arrangement of particles in atoms and molecules. The patterns of electron orbitals, for instance, determine the chemical properties of elements and the behavior of matter.
2. **Macroscopic Patterns**: On larger scales, the wave function's influence is visible in the alignment of particles within crystals, the behavior of quantum fluids like superconductors, and even the collective behavior of biological systems.
3. **Cosmic Patterns**: The large-scale structure of the universe—the distribution of galaxies, dark matter, and cosmic voids—reflects quantum

fluctuations in the early universe, stretched and amplified by cosmic inflation.

These patterns suggest that the universe operates according to deep, underlying principles encoded in the wave function. By studying these patterns, physicists aim to uncover the laws that govern the quantum-to-cosmic continuum.

Philosophical Implications of Observables

The study of observables in a quantum universe raises profound philosophical questions about the nature of reality, measurement, and causality:

1. **Reality and the Observer**: If the wave function represents all possible states, does reality exist independently of observation? The act of measurement collapses the wave function, suggesting that the observer plays a fundamental role in shaping reality.

2. **Determinism vs. Probability**: Classical physics is deterministic, but quantum mechanics introduces intrinsic randomness. Are these probabilities fundamental, or do they reflect deeper, hidden variables?

3. **The Nature of Time**: In a universe governed by the wave function, time may be an emergent property rather than a fundamental dimension. Observables help define a sequence of events, giving rise to the perception of time.

Experimental Insights into Observables

Experimental physics provides critical insights into the nature of observables and the wave function:

1. **The Double-Slit Experiment**: This experiment demonstrates how the act of measurement determines whether a particle behaves as a wave or a particle, highlighting the role of observables in shaping quantum reality.

2. **Bell's Theorem and Entanglement**: Experiments testing Bell's inequalities confirm the non-local correlations predicted by quantum mechanics, challenging classical notions of separability and locality.

3. **CMB Measurements**: Observatories like the Planck satellite and the Atacama Cosmology Telescope study the CMB, extracting information about the universe's quantum origins and inflationary history.

4. **Gravitational Wave Detection**: Observatories like LIGO and Virgo detect and analyze gravitational waves, providing a new way to study the quantum dynamics of massive astrophysical events.

Making Complex Physics Accessible

Explaining the role of observables in a quantum universe requires bridging abstract mathematics and intuitive understanding. Metaphors and analogies can help:

1. **The Wave Function as a Canvas**: Imagine the wave function as a vast canvas of potential realities. Observables act like an artist's brush, selecting and revealing specific patterns from this canvas.
2. **Quantum Measurement as a Coin Toss**: While the wave function provides the probabilities, the act of measurement is like flipping a coin, collapsing the probabilities into a definite outcome.
3. **Cosmic Waves as Echoes**: Cosmic waves, like echoes in a canyon, carry information about the universe's quantum origins, allowing us to reconstruct its history.

By using such analogies, we can make the quantum universe more accessible and relatable, even to those without a background in advanced physics.

Conclusion

Observables are the keys to unlocking the mysteries of the quantum universe. They bridge the abstract probabilities of the wave function with the tangible phenomena we can measure and experience. From the quantum behavior of particles to the large-scale structure of the cosmos, observables reveal the hidden patterns and underlying principles that govern reality.

Cosmic waves, as observables rooted in quantum mechanics, link the microcosmic and macrocosmic scales, providing insights into the universe's earliest moments and its ongoing evolution. By studying these waves and other quantum observables, we continue to crack the code of the wave function, uncovering the fundamental laws that shape our existence.

In this endeavor, the interplay between theory, observation, and philosophy enriches our understanding of the universe, reminding us that even in the realm of probabilities and uncertainty, there is order, beauty, and profound meaning.

The Anthropic Principle and Wave Function Probabilities

The universe is a delicate symphony of constants, laws, and interactions that have conspired to create a cosmos capable of supporting life. From the force of gravity to the charge of the electron, the physical parameters governing the universe appear to be fine-tuned in a way that allows stars, planets, and living organisms to exist. This apparent fine-tuning has long puzzled scientists and philosophers, and its resolution might lie in the quantum wave function—the fundamental mathematical construct that encodes all possible states of a quantum system.

The interplay between the wave function and the anthropic principle provides a fertile ground for exploring profound questions: Why does the universe have the properties it does? How does quantum mechanics influence the conditions for life? And could the multiverse, a speculative extension of quantum theory, explain the fine-tuning we observe? By connecting wave function probabilities to the anthropic principle, cosmic waves, and hidden patterns in the universe, we can begin to unravel these mysteries and illuminate their relevance to our understanding of reality.

The Wave Function and Fine-Tuning

The wave function is a cornerstone of quantum mechanics, describing the probabilities of all possible states of a system. In the context of the universe, the wave function encodes a vast ensemble of potential realities, each with its own set of physical constants, laws, and initial conditions. The version of the universe we observe is one realization of this ensemble, made manifest through quantum measurement and the collapse of the wave function.

Fine-tuning refers to the observation that many physical constants must fall within extremely narrow ranges for life as we know it to exist. For example:

- The cosmological constant, which drives the acceleration of the universe's expansion, must be finely balanced to prevent a collapse or runaway expansion.
- The strong nuclear force must be just strong enough to bind atomic nuclei but not so strong that stars burn out too quickly.
- The ratio of the electron's mass to the proton's mass must allow for stable chemical bonds essential to life.

The wave function provides a framework for understanding this fine-tuning. If the wave function represents a superposition of all possible universes, the one we inhabit is a subset of those where the constants and conditions align to allow for observers. This connection is where the anthropic principle enters the discussion.

The Anthropic Principle

The anthropic principle states that the universe's properties must be compatible with the existence of conscious observers because only such a universe can be observed. This principle comes in two forms:

1. **The Weak Anthropic Principle**: Observers exist only in universes where conditions permit their existence. The observed fine-tuning of the universe is thus a selection effect—if the universe weren't suitable for life, no one would be around to notice.

2. **The Strong Anthropic Principle**: The universe is compelled, in some sense, to develop conditions that allow life and consciousness to emerge.

The wave function connects directly to these ideas by framing fine-tuning as a probabilistic phenomenon. In the ensemble of possible universes described by the wave function, most configurations may not support life. However, the anthropic principle suggests that we are naturally biased to observe one of the rare universes that does.

Quantum Mechanics and the Multiverse

One way to address fine-tuning is through the concept of the multiverse—a collection of universes, each with different physical constants and initial conditions. In this view, the wave function doesn't describe just one universe but a vast landscape of possibilities. Our universe is one bubble in this multiverse, selected by the anthropic principle because it permits observers.

The multiverse is supported by several theoretical frameworks:
• **Inflationary Cosmology**: During the rapid expansion of space known as inflation, quantum fluctuations could have created pockets of spacetime with different properties, giving rise to a multiverse.
• **String Theory**: This theory suggests that different solutions to the equations of

string theory correspond to different universes with varying physical laws.
- **Many-Worlds Interpretation**: In quantum mechanics, this interpretation posits that all possible outcomes of a quantum event occur, each in a separate, branching universe.

While the multiverse is speculative, it provides a natural explanation for fine-tuning: with enough universes, it is inevitable that some will have the right conditions for life.

Cosmic Waves and the Anthropic Principle

Cosmic waves, such as the fluctuations imprinted on the cosmic microwave background (CMB), offer observational insights into the quantum origins of the universe. These waves originated as quantum fluctuations in the early universe, amplified during inflation into macroscopic ripples in spacetime. They contain information about the initial conditions of the universe and the probabilities encoded in the wave function.

By studying cosmic waves, scientists can probe the fine-tuning of the universe. For example:
- The density fluctuations observed in the CMB are directly tied to the formation of galaxies and large-scale structure. Without these fluctuations, the universe would be a uniform, lifeless expanse.
- The polarization patterns in the CMB reflect the influence of quantum fluctuations during inflation, providing clues about the physics of the early universe.

These observations demonstrate how quantum mechanics and wave function probabilities shape the universe's properties, including those that allow for the emergence of life.

Hidden Patterns in the Universe

The wave function encodes not only probabilities but also patterns—subtle structures that connect the quantum and cosmic scales. These patterns emerge in several forms:

1. **Quantum Superposition**: At the quantum level, particles exist in multiple states simultaneously, creating interference patterns that reveal the underlying probabilistic nature of the wave function.
2. **Cosmic Structures**: On the largest scales, the distribution of galaxies, galaxy clusters, and dark matter reflects the imprint of quantum fluctuations stretched by inflation. These structures are the hidden patterns of the universe, written in the language of the wave function.
3. **Mathematical Symmetry**: Many physical laws and constants arise from symmetries, such as those described by group theory. These symmetries are manifestations of the wave function's deeper patterns.

4. **Life's Complexity**: Even the intricate organization of living systems can be seen as an emergent pattern, arising from the fine-tuned constants and probabilistic dynamics of the quantum universe.

Understanding these patterns is akin to cracking a cosmic code. By analyzing how wave function probabilities manifest across scales, we uncover the mechanisms that link quantum mechanics to the emergence of life and complexity.

Making Complex Physics Accessible

Connecting wave function probabilities to the anthropic principle and fine-tuning involves grappling with abstract concepts, but these can be made accessible through analogy and visualization:
- **The Wave Function as a Map**: Imagine the wave function as a map of possible universes. Each point on the map represents a different set of physical

constants, and our universe is a rare dot in a habitable zone.
- **Fine-Tuning as Dial Settings**: Picture the physical constants as dials on a machine. Each dial must be precisely adjusted for the machine (the universe) to work. The wave function represents all possible settings, but only certain configurations allow the machine to operate.
- **The Multiverse as a Lottery**: If every universe is a ticket in a cosmic lottery, the anthropic principle explains why we find ourselves holding a winning ticket—life can only arise in the rare universes that support it.

These analogies help bridge the gap between abstract mathematics and intuitive understanding, making the quantum universe more relatable.

Philosophical Implications

The intersection of the wave function, the anthropic principle, and fine-tuning raises profound philosophical questions:

- **Why Is There Something Rather Than Nothing?**: If the wave function represents all possible universes, why does any particular universe exist, and why does it have properties compatible with life?
- **What Is the Nature of Reality?**: Does the wave function describe an objective reality, or is it merely a tool for calculating probabilities? The multiverse idea further complicates this question, suggesting that reality might encompass countless unseen universes.
- **Purpose vs. Coincidence**: Is the universe fine-tuned for life by design, or is our existence a coincidental byproduct of probabilistic processes? The anthropic principle offers a naturalistic explanation, but it leaves room for philosophical debate.

Observational and Experimental Insights

Advances in observational and experimental techniques provide critical tests for theories linking the wave function to fine-tuning and the multiverse:

- **CMB Studies**: Precise measurements of the CMB's fluctuations and polarization patterns continue to refine our understanding of the early universe and inflation.
- **Gravitational Wave Detection**: Observatories like LIGO and Virgo aim to detect primordial gravitational waves, which would provide direct evidence of quantum fluctuations during inflation.
- **Particle Physics Experiments**: High-energy experiments at facilities like CERN probe the fundamental laws governing the universe, potentially revealing connections between quantum mechanics and cosmology.
- **Exoplanet Surveys**: By studying the prevalence of Earth-like planets and their potential for hosting life, scientists can test hypotheses about the universality of fine-tuning.

These efforts bring us closer to answering some of the most profound questions about the universe and our place within it.

Relevance and Intrigue

The connection between the wave function, the anthropic principle, and fine-tuning is not merely an abstract theoretical pursuit. It addresses questions that are central to our understanding of existence: Why does the universe have the properties it does? Could life arise elsewhere? And what does quantum mechanics reveal about the nature of reality?

Cosmic waves and hidden patterns serve as tangible links between quantum probabilities and cosmic phenomena, providing observational evidence for theories that once seemed purely speculative. By cracking the code of the wave function, we not only uncover the principles governing the cosmos but also deepen our appreciation for the delicate interplay of chance, necessity, and complexity that makes life possible.

Future Directions in Quantum Cosmology

Quantum cosmology sits at the frontier of physics, attempting to unravel the mysteries of the universe by extending the principles of quantum mechanics to the grandest scales imaginable. At its heart is the wave function, a mathematical entity that encodes all possible states of the universe. While physicists have made great strides in understanding how the wave function governs particles and forces at small scales, significant challenges remain in applying it to the cosmos as a whole. What lies ahead in this quest? How might theories like quantum gravity, string theory, and loop quantum gravity unlock deeper insights into the wave function? And how do cosmic waves and the hidden patterns of the universe guide this exploration?

This journey into future directions of quantum cosmology combines cutting-edge theoretical advances, experimental pursuits, and philosophical questions, promising profound implications for our understanding of existence.

The Quest to Unite Quantum Mechanics and Gravity

At the core of quantum cosmology is the need to reconcile quantum mechanics, which governs the microscopic world, with general relativity, which describes gravity and the universe at large scales. These two pillars of modern physics operate in entirely different mathematical frameworks, and their unification is one of the great unsolved problems in science.

Quantum Gravity: A Missing Piece

Quantum gravity seeks to create a theory that describes the behavior of spacetime itself in quantum terms. This is essential for understanding the earliest moments of the universe, where the densities and energies were so high that classical descriptions of spacetime break down. The wave function in this context represents not just particles and fields but the quantum states of spacetime itself.

One approach to quantum gravity is the Wheeler-DeWitt equation, which describes the universe's wave function in a timeless framework. However, it raises deep conceptual questions: What does it mean for the wave function to exist "outside" of time? How does this reconcile with our everyday experience of a flowing timeline?

String Theory and the Wave Function

String theory offers another path toward understanding the wave function of the universe. By postulating that the fundamental building blocks of reality are not particles but tiny vibrating strings, string theory provides a framework where quantum mechanics and gravity coexist. These strings vibrate in higher-dimensional spaces, and their interactions create the particles and forces we observe.

In this framework, the wave function encompasses not just our observable universe but also a vast "landscape" of possible universes, each with different physical laws. This has profound implications for quantum cosmology:

1. **The Multiverse Hypothesis**: String theory suggests that our universe might be one of countless others, each with its own distinct wave function. This raises questions about whether the properties of our universe are fine-tuned for life or simply one of many possibilities realized by chance.

2. **Cosmic Waves in Higher Dimensions**: Vibrations of strings in extra dimensions could manifest as cosmic waves in our observable universe, offering testable predictions through their influence on phenomena like the cosmic microwave background (CMB).

3. **Holography and the Universe's Edge**: The holographic principle, derived from

string theory, proposes that the universe's information content can be encoded on a lower-dimensional boundary. This reshapes our understanding of the wave function, suggesting that what we perceive as a three-dimensional universe could emerge from a deeper, more abstract quantum structure.

Loop Quantum Gravity: Discrete Spacetime

An alternative to string theory is loop quantum gravity (LQG), which proposes that spacetime itself is quantized, consisting of discrete units or "loops." In this framework, the wave function of the universe describes the probabilistic arrangement of these loops, akin to a lattice structure underlying reality.

    1. **The Big Bounce**: Unlike classical cosmology, where the Big Bang represents a singularity, LQG suggests that the universe underwent a "big

bounce," transitioning from a previous contracting phase to the current expansion. The wave function in this context encodes both phases, allowing physicists to explore what came before the Big Bang.

2. **Emergent Geometry**: In LQG, spacetime geometry is not fundamental but emerges from the quantum states of loops. This challenges traditional notions of space and time, suggesting that the wave function encodes not just matter and energy but the fabric of reality itself.

3. **Observable Consequences**: Predictions from LQG, such as deviations from classical spacetime at extremely high energies, might be testable through phenomena like gravitational waves or black hole evaporation.

Cosmic Waves and the Quantum Universe

Cosmic waves serve as a bridge between quantum mechanics and the large-scale universe, providing critical observables that guide the development of quantum cosmology.

1. **Quantum Fluctuations and the CMB**: The cosmic microwave background (CMB) preserves imprints of quantum fluctuations from the early universe. These fluctuations, amplified by cosmic inflation, seeded the formation of galaxies and cosmic structures. By studying the CMB's detailed patterns, physicists gain insights into the universe's initial wave function.

2. **Gravitational Waves as Probes**: Gravitational waves carry information about the most extreme astrophysical events, such as merging black holes and neutron stars. On a cosmological scale, they could reveal traces of quantum gravity effects, offering a window into the wave function's role in shaping spacetime.

3. **Primordial Waves**: Detecting primordial gravitational waves—ripples

generated during inflation—would provide direct evidence of quantum processes at the universe's birth. These waves could reveal whether string theory, LQG, or another framework best describes the early universe's wave function.

Hidden Patterns of the Universe

The wave function is more than a tool for predicting probabilities—it encodes the hidden patterns and structures that define the universe. These patterns emerge at multiple levels:

1. **Quantum Entanglement:** Entanglement links particles across vast distances, suggesting that the wave function operates as a unified whole. In quantum cosmology, this interconnectedness could extend to the universe's large-scale structure, where entangled regions of spacetime influence each other.

2. **Self-Similarity in the Cosmos**: Patterns seen in quantum systems, such as fractal-like structures, might echo in the arrangement of galaxies and cosmic voids. This self-similarity hints at universal principles underlying both quantum and cosmic scales.

3. **Wave Function Symmetries**: The mathematical symmetries of the wave function, such as those described by group theory, govern the laws of physics. Exploring these symmetries could reveal why the universe has the specific properties it does, from the strengths of fundamental forces to the existence of matter itself.

Philosophical Reflections on Quantum Cosmology

Future directions in quantum cosmology raise profound philosophical questions that challenge our understanding of existence:

1. **Reality and Probability**: If the wave function encodes all possible states, what determines the specific universe we experience? Are we merely one realization in a multiverse, or does some deeper principle select our reality?

2. **The Nature of Time**: In many quantum cosmology models, time is emergent rather than fundamental. This suggests that the flow of time we perceive is a product of the wave function's dynamics, reshaping our understanding of causality and change.

3. **The Role of the Observer**: The wave function's collapse upon measurement implies a central role for observers in defining reality. In quantum cosmology, does the universe itself "observe" its own wave function, or is an external observer required?

Experimental Horizons

Advancing quantum cosmology requires experimental breakthroughs that test its theoretical predictions. Some promising directions include:

1. **CMB Polarization Studies**: Experiments like the BICEP and LiteBIRD missions aim to detect primordial gravitational waves through their effects on the CMB's polarization. These waves could confirm inflationary models and shed light on the quantum origins of the universe.

2. **Gravitational Wave Observatories**: Next-generation detectors, such as the Einstein Telescope and LISA, will probe gravitational waves with unprecedented sensitivity, revealing quantum gravity effects and the wave function's influence on spacetime.

3. **High-Energy Physics**: Particle accelerators like the Large Hadron Collider (LHC) and its successors may uncover signatures of quantum gravity or

extra dimensions, providing clues about the wave function's deeper structure.

Making Complex Physics Accessible

Quantum cosmology's concepts are abstract, but they can be made accessible through analogies and storytelling:

1. **The Wave Function as a Map**: Imagine the wave function as a map of all possible journeys the universe could take. Observations reveal specific paths, but the map retains the potential for countless others.

2. **Cosmic Waves as a Symphony**: Just as sound waves create music, cosmic waves create the universe's structure. The wave function serves as the score, guiding the symphony of reality.

3. **Entanglement as Cosmic Threads**: Picture the universe as a vast tapestry, with entangled threads weaving connections across space and time. The

wave function is the loom that shapes this intricate pattern.

Conclusion

The future of quantum cosmology lies in bridging theory, observation, and philosophy to decode the wave function of the universe. Theories like quantum gravity, string theory, and loop quantum gravity offer promising paths, each with its own vision of how the wave function shapes spacetime and matter. Observables like cosmic waves and the CMB provide essential data, connecting abstract mathematics to the physical world.

As we uncover the hidden patterns of the universe, we move closer to understanding the fundamental laws that govern existence. These patterns reveal a cosmos that is not random but deeply ordered, with quantum mechanics and cosmology intertwined in a grand, unified picture.

The quest to crack the code of the wave function is not just a scientific endeavor—it is a journey into the nature of reality itself. It challenges us to rethink space, time, and causality while inspiring awe at the profound beauty of the universe. In this pursuit, we uncover not just answers but new questions, ensuring that the adventure of discovery will continue for generations to come.

About Scott Perdue

Scott Perdue is a dynamic entrepreneur, author, and community leader with a life rooted in faith, family, and service. A devoted Christian, Scott has been married for over 20 years and is the proud father of four children—two girls and two boys. His passion for personal development and spiritual growth is reflected in his prolific writing career, having authored over 100 books, most of which focus on self-help and Christian themes. His books have touched the lives of countless readers seeking guidance on how to lead a fulfilling, faith-centered life.

For over 15 years, Scott has been a dedicated member of GUTS Church, a place he fondly refers to as "It Takes GUTS to Serve the Lord." His service to the church and community extends beyond attendance; he spent six years as a representative for the GUTS Food Bank, where he managed the movement of wholesale goods to help those in need. Scott also led a successful Maximized Manhood study group based on Edwin Cole's teachings, further exemplifying his commitment to fostering spiritual growth among men.

An accomplished entrepreneur, Scott has started and operated over 30 businesses, ranging from pest control to contracting. He is the founder of Universal Bug Man, a pest control service where Scott earned a reputation as a "pest control superhero." His entrepreneurial ventures include Tulsa Furniture Wholesale, Tulsa Auction Spot, and Builderhaus Unlimited, among others. Scott's business acumen extends to the health and wellness industry, where his company HCG Medical helped over 20,000 clients lose weight, generating over $6.5 million in sales in its best year.

Scott Perdue is a man of many talents, driven by his faith and dedication to serving others through his varied enterprises and writing.

Follow Scott Perdue on YouTube, Facebook & Visit UniversalWholesaleStore.com

Published Books by Scott Perdue (Buy Today on Amazon)

Christian Books by Scott Perdue:

Biblical Entrepreneur Leadership: Amplified Leverage Business Skills Book & Workbook

Biblical Men's Leadership Skills: Becoming an Amplified Christian Superstar Book & Workbook

Unleashing Biblical Manhood: Taking Ground Like a Warrior Book & Workbook

Promised Land Leadership: Leading an Army Like Joshua

Wilderness Wisdom of Moses: Timeless Life-Changing Leadership Lessons

Rules of Christianity According to Paul Book & Workbook

Provisional Miracles of Jesus: Provision through Supernatural Means Book & Workbook

Kingdom Money: Unlocking Biblical Secrets to Financial Success

The King's Highway: Lean into Jesus for Accelerated Success

Walk in the Works of the Lord: An Amplified Passion Understanding

God's River: Getting into the Kingdom Family Flow

Forgiven & Unoffendable: The Power of Walking Righteously

God is Real: Knowing the Spirit - A Journey Through Faith, Miracles, and Divine Presence

The Gift of Light: A Journey of Spiritual Growth for Life Expansion

On Fire For Jesus: Bring Plasma Energy to Your Heart Pump

Immortal DNA: Living Forever as an Eternal Spirit

Earth is God's Beach Ball: Celebrate the Legacy of Joyful Living

Faith in the Wilderness: Biblical Lessons for Strength and Spiritual Growth

Living on Purpose: A Comprehensive Guide to a Meaningful and Fulfilling Life

Praying for Others: Unlocking your God-Given Authority to Change Lives

Speaking in Tongues: Snippets of Life Improvement Code

Be Fruitful and Multiply: A Biblical Guide to Family Planning and Takes

Biblical Map of the Garden of Eden: Where does this Mysterious Garden Exist?

Methuselah: The Biblical Legacy of Noah's Grandfather

The Spirit of Jezebel in Modern Times: Acceptance vs Repentance

Love's Crossroads: The Rewards of Suffering for Love

Features of a Great Christian Camp: A Priority Spiritual Foundation

Daily Mercy: A Journey Through God's Grace Every Morning

Self Help Books by Scott Perdue:

You Are the Masterpiece: Center of the Universe Life Experience

Legacy Blueprint: How to Build a Generational Legacy

Accomplishing Greatness: 10 Legendary Skill Sets of Self-Made Millionaires

The Passive Income Playbook: 10 Game Changing Strategies to Build Wealth

Beginners Guide to Investing in the Future: Gain Wealth from Cutting Edge Sectors

Motivation for Creation: Unlocking the Spark Within

Master Productivity: Unlock your Path to Success

10 Step Productivity Plan: A Guide to Increasing Life's Results

Mindset of Productivity: A Defined Focused Journey

Mindful Love: Embracing Self Love Through Mindfulness and Compassion

Mindfulness for Personal Growth: Transform Your Life One Moment at a Time

The Ultimate Guide to Winning Friends and Influencing People: Master Communication

The Human Connection: Unlocking the Secrets to Understanding and Relating to Others

Stress Free Living: Simple Strategies for Modern Life

Mind Switch: Are you Over-Thinking Negative Thoughts?

Mastering Self-Control: Unleashing the Power of Discipline for Success in Every Aspect of Life

Rising From The Ashes: How to Rebuild When Life Falls Apart

Unlocking Secrets to Weight Loss: A Comprehensive Guide to Science, Nutrition and Wellness

Effective Diet Supplements for Weight Loss

The Body Detox Blueprint: 10 Essential Steps to Cleanse, Heal, and Revitalize Your Body

Secret 1000 Calorie Cryogenic Diet

Book Sales Formula: 10 Proven Secrets to SkyRocket Your Book Sales

Learn to Enjoy Reading: Your Ultimate Guide to Loving Books

The Ultimate Blueprint to Comedy: Your Guide to Mastering Humor and Making People Laugh

Decluttering Your Home: Take Control of Your Space, One Step at a Time

Real Estate Needs Observation: Hot to Bring Light to Entropy & Chaos

Business Books by Scott Perdue:

Legendary Business Skills: How to Think like an Entrepreneur

Seal the Deal: Mastering Sales Objections to Close Every Sale

10 Step Marketing Launch: Ultimate Guide for a Business Advertising Start Up

Email Marketing Success: 10 Ways to Master Business Email Advertising Strategy

Controlled Decent: How to Close a Business

How to Start a Business Networking Group: Learn to Organize and Motivate Business Leaders

Negotiate Like an Auctioneer: Mastering the Art of Persuasion and Control

Auction House Blueprint: How to Win Bids and Host a Successful Auction

How to Run an Antique Shop: Restoring Antique Relics to Modern Living

Secondhand Success: A Complete Guide to Running a Profitable Used Furniture Store

The Thrift Store Playbook: How to Build, Manage, and Thrive in the Resale Business

How to Start an RV Park: Your Roadmap to Success

Turn Rapids to Revenue: How to Run a Profitable River Float Business

Science Books by Scott Perdue:

Creation of Your Galactic Record: Big Bang, DNA, Creation of the Universe. Boom!

Symphony of Life: How Human DNA Plays Like Music

Quantum Cosmos: The Wave Function of the Universe

An Astronauts Heavenly Perspective: Planet, Society and Economy

Earth is the Seed of Life: A Geometric Flower of Life

Infinite Plants in Every Seed

Dodecahedron Earth: Exploring the Geometric Key to the Flower of Life

Pangaea Cracked Open: A Pre-Flood World without Oceans

Ancient Cathedral Architecture: A Language Of Semantics Lost in Time

Power Independence: DIY Guide to Building Off-Grid Energy Systems

Harvesting Heaven: The Ultimate Guide to DIY Rainwater Collection Systems

Farming Tactics for the Sahara Desert: Ultimate Gardening Guide for Arid Takes

Easy to Find Herbal Remedies

How to Build Free Energy Lighting: 10 Effective Easy to Build Free Energy Lights

Dynamic Forces: Exploring the Undeniable Power of Movement

Creative Books by Scott Perdue:

Zeppelin Airship Enterprise: The Future of Flight and Travel Reimagined

Ancient Plasma Energy Weapons Revealed: The Lost Technology of Energy Weapons

Echoes of Camelot: Unveiling the Secrets and Legends of the Knights

Secret Treasures of Rome Revealed: Explore the Ancient Architecture of Rome

The Egyptian Ankh: Secrets of Eternal Life and Ancient Wisdom

Giants, Nephilim, and the Legacy of Humanity: From Ancient Myths to Modern Mysteries

Prophecy of the Seven Suns: Exploring Parhelia in Biblical Prophecy

Epic Scavenger Hunt of Machu Picchu

Adventures of Buying an Island: Edge of Your Seat Suspense Thriller Adventure

My Neighbor is an Inventor: A Journey into Wilson's World of Innovation

Adventures of the Zoo Janitor: Growing Responsibility By Excellence

Exile's Genesis: Chronicles of the New Frontier

Relics of Oklahoma: Route 66 Treasure Hunt

The Oklahoma Waterfall Hunt

Creation of Your Galactic Record

Buy Now on Amazon or UniversalWholesaleStore.com

Unlock the Secrets of Your Existence with Creation of Your Galactic Record

Have you ever wondered what you're truly made of? What if your purpose is far grander than you've ever imagined? Creation of Your Galactic Record takes you on a profound journey to uncover your divine role in the universe, revealing the simulation of life and your cosmic significance.

Explore the Mysteries of Your Existence:
• Big Bang or Biblical Story of Creation – Was the universe created by chance, or is there a divine blueprint? Discover how science and spirituality intertwine in shaping reality.
• Your Moment of Conception – From the moment of fertilization, your journey begins. Explore how consciousness ignites in the womb and how your soul selects this life.
• The Baby's Soft Spot – More than just a physical trait, this sacred gateway is believed to be a portal connecting infants to higher realms. Learn how it influences early consciousness.
• Pineal Gland Movie Projector – Your third eye holds incredible power. Understand how it projects visions, dreams, and even past-life memories into your waking reality.
• Your DNA Movie Film Reel – Your genetic code is a living record, storing ancestral memories and divine imprints. Unlock the hidden messages written within you.
• A Uni Verse: One Song – Everything vibrates in harmony. Learn how the cosmos, music, and frequency influence your life's rhythm.
• The Mind's Sky – Your thoughts create your reality. Explore the infinite landscape of your consciousness and how it shapes your journey.
• A Calling to Your Purpose – You weren't born by accident. Discover the signs and synchronicities guiding you toward fulfilling your higher mission.
• Your Galactic Record – Everything you think, say, and do is imprinted on the universe. Learn how to shape your legacy across time and space.
• The Vortex Ascension – Elevate your awareness and vibrate at a higher frequency. Discover how to transcend limitations and access divine enlightenment.

This book is more than just knowledge—it's an activation of your true potential. Step into your divine story and make your mark on the cosmos. Order Creation of Your Galactic Record today and unlock the universe within you!

Buy Now on Amazon or UniversalWholesaleStore.com

Earth is the Seed of Life: A Geometric Flower of Life

Buy Now on Amazon or UniversalWholesaleStore.com

Earth is the Seed of Life – A Transformative Journey into the Origins, Evolution, and Future of Life

Discover the extraordinary story of Earth as the cradle of life in this captivating new book, Earth is the Seed of Life. Spanning scientific theories, spiritual insights, and future possibilities, this book provides a comprehensive look at how our planet nurtures life and civilization.

With 10 thought-provoking chapters, this book begins with the Origins of Life on Earth, tracing the emergence of life from its earliest forms. Explore how Earth's unique conditions—its atmosphere, water, and distance from the sun—make it the perfect host for life. From the intricate ecosystems that balance life's diversity to humanity's crucial role as Earth's stewards, each chapter offers readers a profound understanding of Earth's life-giving power.

Dive into the deeper cycles of life, death, and rebirth that sustain ecological systems and discover humanity's responsibility to preserve our planet. As you move through chapters such as Earth's Seeds of Knowledge and Civilization and Preserving Earth for Future Generations, you'll be inspired to think about our collective role in safeguarding this precious world.

The final chapters look to the stars, investigating the potential of spreading life to other planets and the ongoing search for extraterrestrial life, raising questions about Earth's place in the broader universe.

Perfect for readers passionate about science, nature, and our spiritual connection to the Earth, Earth is the Seed of Life offers fresh insights into how life flourished here—and what the future may hold.

Grab your copy today and embark on a journey through Earth's magnificent role as the seed of life!

Buy Now on Amazon or UniversalWholesaleStore.com

Symphony of Life: How Human DNA Plays Like Music

Buy Now on Amazon or UniversalWholesaleStore.com

Unveil the Symphony Within: Discover How Human DNA Plays Like Music

Are you ready to explore the profound connection between your DNA and the melodies of the universe? Symphony of Life: How Human DNA Plays Like Music invites you on a transformative journey to uncover the harmonies embedded in your very being.

This captivating book reveals the "One-Song" that unites all creation. Discover how the structure of DNA contains a hidden rhythm, where nucleotides resonate as musical notes, creating a unique melody for every individual. Journey through the balance of beauty and chaos within the DNA Helix and learn how transgenic manipulation can disrupt this divine harmony, leading to physical and spiritual dissonance.

But there's hope. Symphony of Life illuminates how faith can restore the melody. Embrace redemption through Christ to flip the chaos side of your DNA into harmony, freeing you from ancestral burdens and physical ailments.

Explore practical steps to remain in the "Melody of Joy," becoming a beacon of light that attracts others to your unique orchestration. Understand your life as an unfolding masterpiece recorded in the Book of Life. Learn to create a "Galactic Record" that resonates with eternity and discover how aligning with God's Word amplifies your eternal light, making you a star in God's creation.

Let Symphony of Life guide you to live as the center of the universe's melody—an orchestra of beauty, joy, and eternal purpose. Order your copy today and let the music of your DNA inspire a life of divine harmony!

Buy Now on Amazon or UniversalWholesaleStore.com

I WOULD LOVE TO HEAR FROM YOU

If you found value in this book, I kindly ask for your review and for you to share your experience with others. Please take a moment to leave a review on Amazon and share how this book has impacted your path to personal growth.

Your feedback not only helps me, but it also guides others in their journey towards change. It's through your support and reviews that my book is able to reach the hands of other readers.

Please take 60 seconds to kindly leave a review on Amazon.

If you reside outside of US, please use the link in your order.

*** All it takes is 60 seconds to make a difference! ***

Join the Scott Perdue Fan Mailing List to stay connected and encouraged! Be the first to receive updates, new releases, special announcements, and exclusive content designed to inspire your journey. Simply sign up on UniversalWholesaleStore.com and become part of a growing community that shares faith, purpose, and encouragement.

Universal Wholesale Store LLC

All rights reserved. No part of this publication may be reproduced, distributed or transmitted in any form or by any means, including photocopying, recording, or other electronic or mechanical methods, without the prior written permission of the publisher, except in the case of brief quotations embodied in critical reviews and certain other noncommercial uses permitted by copyright law.

Limited Liability / Disclaimer of Warranty. While best efforts have been used in preparing this book, the author and publishers make no representations or warranties of any kind and assume no liabilities of any kind with respect to accuracy or completeness of the content and specifically the author nor publisher shall be held liable or responsible to any person or entity with respect to any loss or incidental or consequential damages caused or alleged to have been caused, directly, or indirectly without limitations, by the information or programs contained herein. Furthermore, readers should be aware that the Internet sites listed in this work may have changed or disappeared. This work is sold with the understanding that the advice inside may not be suitable in every situation.

Published by Universal Wholesale Store LLC.

Copyright © 2024 by Scott Perdue

www.ingramcontent.com/pod-product-compliance
Lightning Source LLC
Chambersburg PA
CBHW071550220526
45469CB00003B/964